João Marcos Verly Oliveira da Silva
Maurício Novaes Souza

Organic coffee production

João Marcos Verly Oliveira da Silva
Maurício Novaes Souza

Organic coffee production

Agroecological conservation practices and new technologies available to rural producers

ScienciaScripts

Imprint
Any brand names and product names mentioned in this book are subject to trademark, brand or patent protection and are trademarks or registered trademarks of their respective holders. The use of brand names, product names, common names, trade names, product descriptions etc. even without a particular marking in this work is in no way to be construed to mean that such names may be regarded as unrestricted in respect of trademark and brand protection legislation and could thus be used by anyone.

Cover image: www.ingimage.com

This book is a translation from the original published under ISBN 978-620-2-80826-2.

Publisher:
Sciencia Scripts
is a trademark of
Dodo Books Indian Ocean Ltd., member of the OmniScriptum S.R.L Publishing group
str. A.Russo 15, of. 61, Chisinau-2068, Republic of Moldova Europe
Printed at: see last page
ISBN: 978-620-3-47730-6

THANKS

Initially, I thank God for illuminating my path and for being present in my life.

To my family for having supported me in this journey through Ifes

To my friends at Ifes who helped me in different ways.

I would also like to thank my supervisor Prof. Mauncio Novaes, for always helping me in times of need.

"In the world you will have trouble,
but be of good cheer, I have overcome the
world." - John 16.33

SUMMARY

Considering the economic and ecological crisis faced by coffee farming with the intensive use of agrochemicals and the diversity of existing ecological based models, this course completion work (TCC) aims to present the trajectory of coffee farming in Brazil, weaving an analysis of the main impacts caused by this agricultural *commodity,* having as methodological basis the bibliographical report. The document presents the state of the art of organic coffee farming and its potential for the conservation and preservation of biodiversity, rescue of traditional cultivation practices and a reconstruction of agro-ecosystems through the redesign of production units. It also presents the main assumptions of agro-ecological transition for the promotion of sustainable coffee growing: ecological management of soil, pests and diseases, and redesign of agro-ecosystems with agroforestry systems (SAFs) and Permaculture. It was verified that, contrary to conventional coffee farming, responsible for the devastation of extensive forest areas and the contamination of water and soil resources, the production models of sustainable coffee farming can contribute to the conservation of biodiversity and agro-biodiversity, of traditional culture, enabling the perpetuation of family coffee farming, in an approach that transcends coffee production and encompasses greater aspirations, such as the social reproduction of families in rural areas, the quality of life of farmers and the preservation of natural resources for future generations. The TCC also presents the theoretical and practical assumptions for the construction of sustainable agriculture, providing basic subsidies for agro-ecological transition. Emphasis will be given to coffee farming conducted in SAFs, permaculture and highlights the practices of environmental education as a tool to achieve this objective.

Key-words: coffee-growing, traditional agriculture, agro-ecological production bases, biodiversity, sustainability.

ABSTRACT

Considering the economic and ecological crisis faced by agrochemical intensive coffee growing and the diversity of existing ecological-based models, this article aims to present the trajectory of coffee growing in Brazil, an analysis of the main impacts caused by this agricultural commodity, based on a bibliographic report. The article presents the state of the art of organic coffee farming and its potential for conservation and preservation of biodiversity, rescue of traditional cultivation practices and a reconstruction of agroecosystems through the redesign of production units. In addition, it presents the main assumptions of the agroecological transition for the promotion of sustainable coffee growing: ecological management of soil, pests and diseases, and redesign agroecosystems with agroforestry. Contrary to conventional coffee growing, it has been found to be responsible for the devastation of extensive forest areas and contamination of water and soil resources, models of sustainable coffee styles can contribute to the conservation of biodiversity and agrobiodiversity, traditional and family culture, enabling the perpetuation of family coffee farming, from a perspective that transcends coffee production and harbors greater aspirations, such as the social reproduction of families in the rural environment, the quality of life of farmers and the preservation of natural resources for future generations. The article also presents the theoretical and practical assumptions for the construction of a sustainable agriculture, providing elementary subsidies to the agro-ecological transition.

Keywords: coffee farming, traditional agriculture, agroecological production base, biodiversity, sustainability

Table of Contents

GENERAL INTRODUCTION

When we talk about the study of coffee in Brazil or in any other country that produces it in the world, and that has in it one of its main sources of resources, this subject needs to be treated as something current - since the cultivation of coffee is still a source of income for many families, essentially on small rural properties.

Coffee has been present in Brazil since the colonial period and has seen changes in the structure of society, from the emergence of urban life to changes in labour relations. In the period when European immigrants took over the cultivation activity, coffee plantations emerged. At this time, there was also the transition to capitalism, demanding a new division of labour. When this fact reached the economic context of the country, it drove the rapid dispersion of coffee culture (FERRAO et al., 2017).

According to these same authors, at the end of the 19th century, coffee already accounted for 80% of national revenue and remained so until the beginning of the 20th century, sustaining the political and administrative apparatus of the republican regime and providing resources for the installation of the national industrial park. Coffee has been a lever for the economy throughout history, since it provided a capital surplus that led to investment in agricultural and non-agricultural sectors.

Campo (1983) reported:

> "Colombia emerged in the second half of the 19th century as an important coffee country, coinciding with a phase of growing imbalance in the world coffee market. This coffee expansion was also a process of migration from the coffee frontier to the interior of the country".

The trajectory of coffee in Brazil is important not only for the resources generated throughout history, but for generating changes in sectors that go beyond the economy, for example, the changes made in the exchange policy and the standardization of commercialized products.

In Brazil, since the colonial period, the adopted cultivation model has been characterised by monocultures in full sun, with a low level of biological diversity. The conventional model is, for Theodoro (2002), defined as a model that abuses agrotoxics and chemical fertilizers. In terms of organisation, Prado Junior (1967) and Souza (2015) point out that coffee plantations follow the traditional patterns of agriculture in the country, with large-scale exploitation in large plantations, and extensive areas of monoculture. Dean (1997) says that the natural characteristics of the coffee plantation soil were disregarded, opting for the expansion of monoculture aiming at the quantity produced in detriment of quality.

Souza (2006) states that with the predominance of monoculture in an extensive system, the coffee plantations despised the original form of plantation, aging earlier. The coffee tree starts producing at the age of three years, lasting about twenty years, when it loses its productive potential.

Given the high degree of agrotoxic and input use, the search for more natural products has grown considerably, along with movements towards sustainable development. The search for foods from more sustainable production systems, such as organic production methods, is a trend that is becoming stronger worldwide. The way in which the consumer has incorporated organic products to his table has grown and almost reaches the conventional product in some countries, demanding attributes such as price, diversity, regularity.

According to Blackwell et al. (2011), the market considers how consumers respond to behavioural issues, which shows their self-guidance, that is, the goals and behaviours to which they aspire.

Organic agriculture is based on offering healthy food, which guarantees the preservation of the environment, the product and the consumer. Organic agriculture represents much more than the substitution of chemical fertilizers and agrotoxics for products allowed by legislation. This practice demands a constant analysis of production and commercialisation, because such practices need to contribute to the sustainable development of the community as a whole.

According to the Technical Regulations for Organic Production (2011, p. 10), sustainable development means:

> Natural resources are preserved for use by future generations; production generates satisfactory economic results for all people involved; people involved in production and marketing have quality of life and well-being; farmers' culture, way of life and knowledge are respected.

The product is considered organic, besides not using chemical products, when it is cultivated in an environment that takes into account social, economic and environmental sustainability, always valuing the culture of the communities where it is or will be inserted.

To produce organic food, the producer must follow national norms, dictated by the Ministry of Agriculture, Livestock and Supply, in accordance with normative instructions and international norms, in accordance with the International Federation of Organic Agriculture Movements (IFOAM), whether the food is of animal or plant origin.

There are products of authorised use, provided they are generated in the area itself and free of contaminants, such as (BRASIL, 1999):

J Organic compounds derived from organic waste;
J *Manure* (solid or liquid, crop residues such as maize straw, green manure by planting certain species of plants, preferably species belonging to the leguminous, gramineous, cruciferous or cereal families);
J Biofertilisers, which can be obtained from the anaerobic fermentation of crop residues or animal manure in the production of biogas; and
J Beneficial micro-organisms or enzymes, provided they are not GMO/transgenic.

Products from outside the production system must be authorised by a certifier, and these (BRASIL, 1999):

J Vermicompost - organic compost from worm composting;
J From organic waste;
J Composted or liquid manure;
J Vegetable Biomass (oil fermentation);
J Some industrial wastes such as horns, blood, bone dust, hair and feathers, cakes, vinegar and the like as fertiliser supplements;
J Seaweed and its derivatives, and other products of marine origin;
J Fish and fish products;
J Rock or sawdust, shells and derivatives, not contaminated by preservatives;

J Microorganisms, amino acids and enzymes, provided they are not GMO/transgenic;

J Ash and charcoal;

J Biofertilisers;

J Clays or vermiculite;

J Urban composting, as long as it is not directly used on plants and soil, comes from selective collection and is proven to be free of toxic substances;

J Carbonate, as a source of micronutrients.

Thus, the objective of this work is to present the trajectory of coffee farming in Brazil, weaving an analysis of the main impacts caused by this agricultural *commodity,* based on the bibliographical report methodology.

Also, new options for the implantation and management of a coffee plantation will be presented, the production systems: organic and agro-ecological.

CHAPTER I

ENVIRONMENTAL EDUCATION IN COFFEE FARMING

Joao Marcos Verly De Oliveira Da Silva [1]Mauricio Novaes Souza [2]
Fabio Gomes Zampieri [3]
,,.. 3
Stephan Lopes Carvalho

1. IntroductioN

Brazil regulates Environmental Education by the National Policy of Environmental Education (PNEA), which defines the basic principles of Environmental Education in education systems (Law 9.795, of April 27, 1999) (TOTE; ANDRADE, 2015).

Environmental Education is a dimension of education that presents a critical character and that modifies the form of action, collective attitude or collective ideal, in addition to respecting society (MAIA, 2014). Nowadays, there is still an innocent view of Environmental Education that has a greater purpose, but its coverage goes far beyond, since it promotes the link between ecology and society, also involving local, regional and global knowledge (CARVALHO, 2004b).

The educational process carried out in society and the different forms of understanding of the socio-environmental relationship (strict sense) that are the basis of the environmental movement and other social movements do not allow us to define a single Environmental Education (CARVALHO, 2002).

It makes to know that the Environmental Education is formed by several conceptions, with particular paradigmatic visions of nature and of society, the seduction for contrary concepts of a group and interpretations in permanent conflict and dialogue. Within this educational field occurs the encounter of environmental education with critical thinking (CARVALHO, 2004a).

Critical thinking, from the perspective of Environmental Education, assists in the formation of individual-society and collective relationships. Critical Environmental Education assumes in advance a position of responsibility for the world, it must be responsibility with the environment and with the relations of others with the environment, there is no division or order among the dimensions of human support (CARVALHO, 2004a).

---

1 Bachelor in Coffee Technology. Ifes - Campus de Alegre, Caixa Postal 47, CEP: 29500-000, Alegre-ES. jmverly@gmail.com
2 Dr. Professor of the Instituto Federal do Espirito Santo - Campus de Alegre, Caixa Postal 47, CEP: 29500000, Alegre-ES. mauricios.novaes@ifes.edu.br
3 Masters student in Agroecology in the Post-Graduation Programme in Agroecology of Ifes - Campus de Alegre, Caixa Postal 47, CEP: 29500-000, Alegre-ES. fgzampieri@hotmail.com

Maia (2015) highlights that Environmental Education has a critical panorama that suggests an explanation of political ideas, investigating in a thorough way, and the support to overcome the capitalist force, resisting, consequently, a centralized conversation, in the ecological scenario which suggests a visible change, and not social restructuring.

In this sequence, the debate on less aggressive ways of food production in the context of Environmental Education, demonstrates a new social and environmental look at the relations of the worker and his production with the consumer.

2. Environmental education and the promotion of sustainability

The occupation of natural spaces by human beings has caused negative impacts and environmental degradation. The agricultural practices used in conventional coffee farming have been promoting an accelerated process of degradation. On the other hand, organic coffee farming is conceived under the rules of organic agriculture, which aims to strengthen biological processes through crop diversification, fertilization with organic fertilizers and biological pest control, without generating impact on the environment and consumers.

The growing concern of society with health, quality of life and of the environment leads consumers to value the adoption of agricultural production methods that guarantee the quality of products and that are less aggressive to the environment and socially fair to rural workers. It is in this context that organic agriculture emerges as an alternative for more sustainable, environmentally balanced and socially just agricultural production (Figure 1).

It is necessary to understand the negative environmental impacts that the planet is going through and to try to raise awareness of sustainable production, which is one of the ways of dealing with the social and environmental imbalance suffered.

Figure 1. organic and agro-ecological coffee farm

Source: Personal archive.

One of the tools to reverse this process is the environmental education, since it will allow the growth of awareness and sensitivity of the human being, in relation to the preservation of environmental issues.

In general, the lack of awareness about the use of fertilizers and agrotoxics in coffee farming after the Green Revolution, which is ingrained in farmers, has led to the contamination of the environment, endangering the health of farmers and the people around them.

In view of this problem, it is necessary to think of public policies, creating norms and methods that will guide the rural extension activities, providing guidance through environmental education, as a proposal for sustainable development and guaranteeing social and environmental well-being (ANJOS, 2009).

Thus, the environmental education has a role in the foundation of concepts, in the awareness of the current coffee producers, who are reflexive about the ecological form of the new coffee culture, in terms of awareness, preservation and protection of the environment.

As stated by Dantas (2017), environmental education was born with the aim of generating an ecological awareness in each human being, concerned with providing the opportunity for knowledge that would allow a change in behaviour aimed at protecting nature.

According to Barchi (2016):

> "The institutionalization of environmental education has as one of its main justifications the fact that without it it is not possible to create sustainable and just societies, much less guarantee a healthier and cleaner planet for future generations.

The agricultural practices adopted in organic coffee farming encourage the preservation of the ecosystem, and reinforce practices focused on sustainability through environmental education.

According to Souza (2018; 2021), environmental sustainability defines the way in which human beings use goods and natural resources to meet our needs, without depleting them and providing for the next generations.

It is simple: be sustainable and use and take care so that the next one that is going to use it does not lack, thus forming a solidary chain that seeks to preserve the environment in the best possible way.

To create and develop new methods that ensure sustainability within the economic growth, thus encompassing sustainable development, is a challenge that must be put into practice in the daily life of the coffee producer. However, the great beneficiary of adopting good production practices that lead to sustainability is the rural producer himself and his property.

3. Sustainability

The principle of sustainability arises with globalization. Environmental sustainability is the conservation of the environment: for this to be effective, it is necessary to involve rural producers in the search for an agriculture with quality of life. Environmental education is the scientific basis for sustainability, which is a process that should reach society as a whole, finding ways of development that are able to maintain the standard of living of current generations, without compromising future ones. This integration is fundamental so that, finally, development occurs from a sustainable production base.

Sustainability is a process which must be established over the long term. It means that growth and meeting human needs can be achieved without compromising future generations by the indiscriminate exploitation of natural resources. From
according to Leff (2001):

> "The principle of sustainability appears as a response to the fracture of the modernizing rationality and as a condition to build a new productive rationality, based on the ecological potential and on new senses of civilization from the cultural diversity of the human race. It is about the re-appropriation of nature and the invention of the world; not only of a world in which many worlds fit, but of a world made up of a diversity of worlds, opening the encirclement of the globalized economic-ecological order.

These concepts provide the theoretical basis to reach sustainability. It is by integrating the political, social, economic and environmental spheres that we will have the totality of sustainable development, which has given subsidy to create a sustainability curriculum for coffee.

4. Coffee Sustainability Curriculum

According to Esteves (2015), the Coffee Sustainability Curriculum, its format and curricular structure, was based on the Normative Instruction 49/2013, of the Ministry of Agriculture and Livestock - MAPA. The principles of coffee sustainability were created by different institutions that determined technical criteria of production, on good agricultural practices and management of the coffee activity.

The Global Coffee Platform (GCP) is a global initiative with the objective of increasing the use of sustainable practices in coffee production, which is helping to disseminate and apply the Coffee Sustainability Curriculum (CSC) to guide producers in the application of sustainable practices in coffee production.

The CSC indicates what can be done and what should be avoided, with conduct divided into 18 thematic areas, covering social, environmental and economic aspects.

> Economic Aspects:

Its objective is to improve the application of resources and reduce costs. It considers the following aspects: 1. productivity; 2. controls, records and

Soil analysis, fertilizer plan and foliar analysis; 5. Integrated Management of Pests and Diseases.

> **Environmental Aspects:**

Its objective is to meet preservation standards and reduce environmental impacts. It considers the following aspects: 6. coverage and conservation of soil; 7. APP: Permanent Preservation Area; 8. rational use of water; 9. treatment and disposal of waste; 10. Storage of agrochemicals; 11. Return of agrochemical containers; 12. Agrochemicals with registration and grace period; 13. Climate.

> **Social Aspects:**

Aims to meet human needs while maintaining social and environmental safety. It considers the following aspects: 14. IPE use; 15. Training; 16. Health and safety; 17. Labour legislation; and 18. Youth, women and family succession.

Figure 2 shows a farm showing the eighteen (18) thematic areas to guide farmers in applying sustainable practices that encompass social, environmental and economic aspects.

Figure 2. farmland presenting the 18 thematic areas to guide farmers in applying sustainable

practices

5. Implementation of the Coffee Sustainability Curriculum (CSC)

With the implementation of the Coffee Sustainability Curriculum (CSC), the benefits generated are:

The producer who achieves good compliance with this content will be more sustainable;

Sustainable producers make more profit: in the short, medium and long term;

J Efficient property management;

J Saves inputs, increasing profitability;

J It improves coffee yield and quality;

J Greater control of production costs;

J Soil and water is preserved and productive impacts mitigated and/or ended;

J Focus on capacity building and training (farmer encouraged to do training);

J Better enforcement of legislation;

J Greater security to work; and

J Better organisation of the property.

All the benefits described in coffee production aim to maintain its superior quality, without leaving aside the social, environmental and economic concerns of the producer.

6. Concluding remarks

The technological innovations that have occurred in the last decades have significantly amplified the yield of crops and improved the work conducts in the field. The increase in productivity was remarkable due to the introduction of capital goods, inputs and new technologies that the industry has made available to the market. However, it has brought negative side effects due to errors and exaggerations, which eventually cause damage to consumers, farmers, society and the environment.

This development model has been the producer of underdevelopment. It is worth remembering that poverty is incompatible with balance and environmental sustainability: if this process is not stopped, degradation will persist. For this reason, it is necessary to create a new consciousness in society, where ethical principles are developed, in order to make a real effort to overcome the current planetary crisis.

CHAPTER II

LEGISLATION FOR ORGANIC PRODUCTION

Joao Marcos Verly De Oliveira Da Silva[4] Mauricio Novaes Souza [5]Fabio Gomes Zampieri[6]

4 Bachelor in Coffee Technology. Ifes - Campus de Alegre, Caixa Postal 47, CEP: 29500-000, Alegre-ES. jmverly@gmail.com

5 Dr. Professor of the Instituto Federal do Espirito Santo - Campus de Alegre, Caixa Postal 47, CEP: 29500000, Alegre-ES. mauricios.novaes@ifes.edu.br

6 Master's student in Agroecology in the Post-Graduation Programme in Agroecology of Ifes - Campus de Alegre,

Mauricio Lorengao Fornazier4

1. Introduction

Organic agriculture is regulated by laws and decrees. Law No. 10.831 of 23 December 2003, regulates the inclusion of production, storage, labelling, transport, certification, marketing and inspection of organic products. Decree No. 6323 of 27 December 2007 creates the Brazilian system for assessing organic conformity. This decree is composed by the Ministry of Agriculture, inspection bodies of the States and organic conformity assessment bodies.

There are agrotoxics used which have in their composition products allowed by the organic legislation and which are regulated.

Decree n°4074/2002 states:

> "Phytosanitary product with approved use for organic agriculture - agrotoxic or similar, containing exclusively substances permitted, in specific regulations, for use in organic agriculture [...] reference specification - minimum specifications and guarantees that phytosanitary products with approved use in organic agriculture must follow to obtain registration".

After registration, these products are called phytosanitaries and are considered of low environmental impact and toxicity. Some examples:

J **Alternative products against fungal diseases:**

- Simple sulphur and its preparations, at the discretion of the certifier;

- Stone powder [1/3 aluminium sulphate and two thirds clay (kaolin or bentonite) in 1% solution];
- Copper salts, in fruit growing;
- Propolis;
- Hydrated lime, only as fungicide;
- Iodine;
- Plant extracts;
- Compound and plant extracts;
- Vermicompost;
- Bordeaux mixture and sulphocalcic lime, at the certifier's discretion; and
- Homeopathy.

J **Means against insects and pest pathogens:**

- Viral, fungal and bacteriological preparations that are GMO/transgenic (only with the

Caixa Postal 47, CEP: 29500-000, Alegre-ES. fgzampieri@hotmail.com; mauzier_if@hotmail.com specific permission of the certifier);

- Insect extracts;
- Plant extracts;
- Oil emulsions (without chemical-synthetic insecticides);
- Soap of natural origin;
- Coffee powder;
- Gelatine;
- Rock dust; and
- Ethyl alcohol.

In order for organic food to be produced and reach the market, it is necessary to comply with some determinations based on Brazilian legislation.

Law No. 10.831 of 23 December 2003 states that the organic farming production system is characterised by a production process that maintains the natural conditions of the environment.

According to the criteria established by the above-mentioned law, to be commercialised, organic products must be in accordance with certification by officially recognised organisations.

Decree No. 6.323, 27 December 2007, provides that all organic products must have/be:

- ANVISA inspection;
- official accreditation;
- organic certification;
- Organ integrity;
- social control organisation;
- go through the conversion period;
- organic quality;
- organic quality assurance;
- in addition to guaranteeing respect for tradition and culture and the mechanism for social organisation of traditional communities, among others.

2. Organic coffee

Among the organic products cultivated around the world, coffee is one of the most important, both in the national and international economy, being the second most consumed beverage in the world. As a consequence of this fact, the production areas of conventional cultivation have grown a lot, which causes negative environmental impacts. In this context, the need for awareness-raising as a tool to curb aggression arises (ZACARIAS et al., 2019).

In this sense, Sirvinskas (2008) reports:

> In order to protect the environment, it is necessary to make people aware of the relationship between man *and the* environment (SIRVINSKAS, 2008).

To better control and preserve the soil, organic coffee cultivation emerges, which is grown without the use of synthetic fertilizers and/or genetically modified organisms. The production and commercialisation of organic coffee go through evaluation stages in accordance with technical norms and standards.

Brazil has laws that establish certifications for cultivation and commercialisation. This is the case of the law on organic agriculture N° 10831/2003 which states in its article 3:

> To be marketed, organic products must be certified by an officially recognized organism, according to criteria established by regulation. (BRAZIL, 2003).

Agriculture can cause potential environmental problems, like any other human activity. Coffee farming is no different, causing negative environmental impacts if practised in disharmony with the rules of conservation and/or preservation of the environment.

It is up to the coffee grower to exercise his environmental citizenship, walking together with sustainability. In the case of coffee cultivation, the question of thinking only about production for profit is increasingly questioned, since the effects of inputs on the environment are lasting and / or irreversible, existing the need to adopt sustainable practices.

According to Lima (2002), soil conservation practices, ecological management of pests and diseases, invasive plants and, more specifically, the correct disposal of effluents resulting from the pulping of coffee are indispensable.

In the case of organic coffee, the production is made without agrochemicals and fertilizers of high solubility, or that can be substituted by complex by-products of recycling of vegetal and/or animal organic matter, animal manure, coffee pulp or husk, among others. However, Theodoro (2003) points out that there are disadvantages to this model of cultivation, since this production yields a higher productivity than traditional coffee.

Graziano et al. (2009) and Wachsner (2005) emphasize that each organic product presents in its package the stamp of certification. This will serve as an instrument of guarantee to the buyer that he is consuming the result of a production system that aims at not damaging the environment, that adopts practices of conservation and recuperation of environmental diversity, aiming at sustainability and food safety. It is also necessary to consider the respect for labour regulations, indispensable for organic certification. It is an alternative way of offering the transparency demanded by consumers.

A large part of the organic coffee plantations have their origin in conventional coffee plantations, which went through a conversion process of, on average, three years without the use of defensives and synthetic fertilizers.

Case Studies" will be presented which represent new models of organic coffee production and the new technologies available to the farmer, following sustainable proposals of production and management.

CHAPTER III

PERMACULTURE IN AGRICULTURE

Joao Marcos Verly De Oliveira Da Silva [7]Mauricio Novaes Souza [8]Otacllio Jose Passos Rangel4 Lucyelen Costa[9] Mauricio Lorengao Fornazier5 Geisa Correa Louback5 Nagila Scarpi Nespoli5 Grazielli Pirovani5

1. Introduction

The "Permaculture" movement developed in Australia from the idea of creating sustainable agro-ecosystems by simulating natural ecosystems, giving priority to perennial crops as central elements (KHATOUNIAN, 2001).

Castro (2014), in his basic guide on permaculture, defines this practice, also called "permanent culture", as a series of concepts and practices focused on sustainability in environmental, cultural, social and economic aspects. Permaculture is grounded in ethical principles that can be used to establish, design, coordinate and enhance all efforts by individuals, places and communities working towards a sustainable future.

2. Permaculture and sustainable development

Permaculture, like Agroecology, is not only an agricultural production model that seeks sustainable rural development. These two sciences seek to encompass social, cultural, historical and ethical aspects, being concerned with the redesign of systems where local resources are valued, energy sources and a production that does not seek quantity and high values in products, but rather values the quality and all the history involved,

7 Bachelor in Coffee Technology. Ifes - Campus de Alegre, Caixa Postal 47, CEP: 29500-000, Alegre-ES. jmverly@gmail.com

3 Dr. Professor of the Instituto Federal do Espirito Santo - Campus de Alegre, Caixa Postal 47, CEP: 29500000, Alegre-ES. mauricios.novaes@ifes.edu.br; ojprangel@ifes.edu.br
8 Master's Degree student in Agroecology at the Post-Graduation Programme in Agroecology of Ifes - Campus de Alegre, Caixa Postal 47, CEP: 29500-000, Alegre-ES. fgzampieri@hotmail.com; mauzier_if@hotmail.com

considering not only scientific knowledge, but also traditional knowledge that collaborates to the development of an agriculture increasingly respectful of the limits of nature (HARLAND, 2018; SOUZA, 2018).

Organic family farming, by presenting itself as a productive system which aims at the self-sustainability of agricultural property, the supply of healthy food and the preservation of environmental and social health, questions the negative repercussions of the modern food production system and comes close to the notion of quality of life.

To better illustrate the link between quality of life and organic family farming, we sought to know the repercussions of the adoption of an organic production system on the quality of life of family farmers from the Association of Farmers of the Encostas de Serra Geral (AGRECO), in Santa Rosa de Lima, SC (AZEVEDO et al., 2011).

This field study helped to elucidate the complexity of the concept of quality of life in rural areas. At the same time, it highlighted the practice of organic family farming as an effective strategy to promote quality of life and social values in this environment and allowed us to define, with greater certainty, the relationship between the proposed categories - quality of life and organic family farming. Consequently, "organic family farming becomes an instrument for the promotion of social values and quality of life in rural areas, with equally important repercussions on the quality of life in urban areas" (AZEVEDO et al., 2011).

3. Permacultural principles

Mollison (1999) points out the construction of the permaculture space, approaching the principles of *design,* where permaculture is not seen as the landscape itself, nor even the skills of organic farming and sustainable agriculture, but more than that, it aims to plan and establish, manage and improve these and all other components, families and communities towards a sustainable future (Holmgren, 2013).

Permaculture *design* is seen as planning, also important in planning for coffee production, not only for production, but also for social and environmental purposes. In permaculture this dynamic planning deals with the relationship between the various elements that are part of the system (plants, buildings, animals, and infrastructure). Figure 1 shows a bio-construction and vegetable planting project.

Figure 1. Bioconstruction and vegetable planting

Source: http://revistaaerea.com.br.

As a direct result of the implementation of permacultural methods, the aim is "the harmonious integration between people and landscape, providing their food, energy, housing and other materials and non-materials in a sustainable way" (mollison, 1999). (MOLLISON, 1999).

Mollison (1999) presents the *design of* the agro-ecosystem as primordial for the success of production and for the environmental, social and cultural benefits. He presents in one of his works, in 1998, a diagram demonstrating the design explained (Figure 2).

Local components such as soil, water and plants must be interconnected with abstract components such as time to prepare the area for cultivation, the necessary data and production ethics.

Furthermore, the social (people and legal support) and energy (technology and structure) components must also be linked to production. The focus is therefore on the design of the production area and all aspects and components involved.

Figure 2. Permaculture *design* diagrams

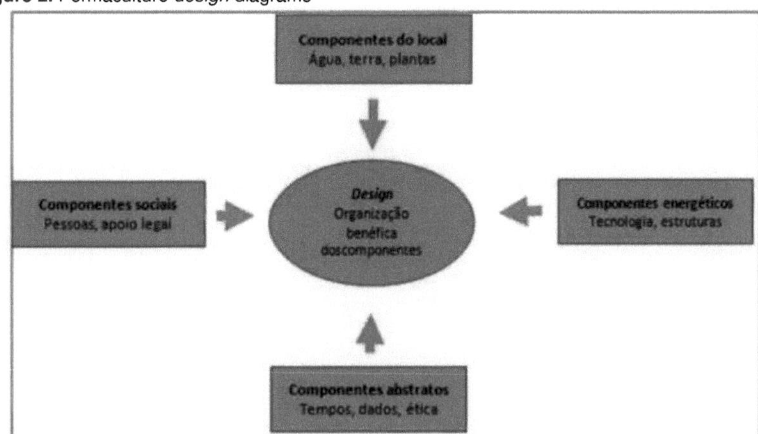

Source: Mollison, 1998.

Mollison and Slay (1998) describing the idea of working with nature, saving energy, result in guidelines to ensure a good *design* system, summarised in Table 1, with applications to coffee farming.

Table 1 - Proposed coffee application guidelines

Guideline	Features	Coffee applications
Relative location	Elements designed in such a way that a mutual relationship is established	Use of aromatic species in tillage
Multiple fungi	All functions should be valued	Spontaneous plants with allelopathic effects
Multiple elements	Various elements responsible for water and energy production and waste use	Residual water from coffee de-pulping
Using biological resources	Replacing toxic elements such as herbicides and pesticides	Green manure and biological control

Diversity	The more diverse the system, the more balanced and sustainable it is	Consortium and biodiversity
Local Energy Recycling	Cycles that concentrate energy efficiently	Water catchment using the topography of the area
Small-scale intensive system	Making the most of the space	Maintain vegetation cover throughout the area
Set and maximise the Borders	The agglutinated borders become another ecosystem	Use of windbreaks and other tree species
Natural succession	Modification processes and natural evolution	Growth of spontaneous plants, ensuring diversity
Attitude	Ability to see and learn from nature	See insects as indicators, not pests
Efficient energy planning	Permaculture zoning of the areas	Land division, age and crop production

Source: Adapted from ollison and Slay (1998).

In permaculture, the techniques and technologies used involve low-cost materials available locally, which can enable the production of coffee by small producers who cannot afford to invest in external inputs. The application of these techniques suggests the reintegration between man and nature, involves creativity, free work and self-management (MOLLISON *apud* Oliveira et al., 2017).

4. Planning and coffee growing in a permaculture system

The permaculture agricultural system is planned so that the system itself can provide the energy necessary for maintenance, stimulating diversity and the production of food, guaranteeing food and nutritional security.

Thus, according to Oliveira et al. (2017), in permaculture soil management can be based on techniques from several alternative models combined.

Figures 3 and 4 show the coffee plantation of the producer Iuri Tristao, from Brejetuba, the main arabica coffee producing municipality in the State of Espirito Santo, receiving most of the wastewater from coffee processing (IRRIGAQAO, 2015).

Your statement:

> "We started to adequately dispose of the water from coffee processing after the environmental license was granted in 2010. Through analysis, we saw that the soil became richer in potassium. This year, we have started to plant corn and yam crops in this area. The prospect is to save on fertiliser, since the soil is already enriched with this nutrient, potassium".

Figures 3 and 4: Irrigation system with wastewater, Brejetuba-ES

Source: Irrigapao (2015).

In the case of coffee farming, for a long time the activity was seen and carried out in the form of monoculture, or in other words, only the production of coffee, based on the conventional production system, where the use of external inputs, agrochemicals, machinery, among others, degraded and put at risk the environmental quality (Figure 5).

Figure 5: Degraded area of coffee cultivation in the conventional production system

Source: Souza (2018).

5. The evolution of permaculture and its principles

With the emergence of fronts organised in favour of environmental quality and the strengthening of family farming, there have been several agricultural production models that seek sustainability in rural areas, strengthening groups, valuing local inputs and ensuring food security (SOUZA, 2018).

Permaculture, besides being a holistic science of socio-environmental nature, becomes a philosophy of life where thinking and living consider that human beings are an integral part of nature and not its dominators. In this context, the human being enters as an actor who promotes the processes of natural interactions favoring and enhancing them, with a view to the production of food, energy, water and other items necessary for human life (SANTOS, 2014).

Adopted in many parts of the world as a planning tool for rural units, permaculture seeks to improve efficiency in their management, seeking to improve the use of natural resources, as well as people's energy consumption. In practice, over time, permaculture becomes incorporated into the lives of those who adopt it, changing their aspirations and ways of seeing what quality of life really is and that it can be achieved together with nature.

In the advance of wellbeing, generally a scenario is installed where self-sufficiency stimulates the permanence of people in the countryside and, in many cases, also human migration from urban to rural environments.

From practical experiences, the concept of permaculture has continued to develop over the last four decades. Currently, it is based on three ethical principles, being the main basis for all work to be done. The ethics of permaculture involve three major needs regarding coexistence with the planet Earth, which basically express their concern with the environmental, social and economic issues (Holmgren, 2013). Thus, these three ethics constitute the vertices of the triangle of sustainability, where its bisector is the point of balance to be reached (Figure 6).

In a second ambit are placed the twelve planning principles which help in the direction of support. The twelve principles of permaculture planning were developed over more than two decades and published in 2002 by David Holmgren in his book "Permaculture: principles and paths beyond sustainability", published in Portuguese in Brazil in 2013.

Figure 6. ethics (central) and planning principles (peripheral) of permaculture

Source: Permacolective (2008), proposed by Holmgren (2013); Dixon and Spotten (2014); WithOnePlanet (2017) and Harland (2018).

Figure 6 presented the systematisation of all the planning principles.

The first ethic, "Care for the Earth", considers the environmental issue and refers us to the obligation to care for the Planet with the awareness that the Earth is everyone's home, helping us to understand that everything and everyone is interconnected and we are interdependent (HOLMGREN, 2013).

The second, "Caring for People", defines a line of action for social issues and seeks to alert to the needs of each person: physical and mental health, quality of life, political autonomy, housing, food security, human rights and education (Holmgren, 2013).

The third, "Caring for the future" (DIXON; SPOTTEN, 2014; WITHONEPLANET, 2017; HARLAND, 2018), is based on the assumption of abundance, where everyone shares the surplus they have and the responsibilities for what they do not have, so that no one lacks anything. In addition, it works on issues of home management, which involves the recognition of limits to growth, including population growth, providing for its stabilization in order to be in line with the carrying capacity of the planet.

In the ethical umbrellas are twelve principles of spatial/territorial planning that work on the interaction between the human species and nature, through the management of natural resources such as energy and water, the use of natural evolutionary processes as inspiration to establish in space and time an environment that can house the human being within their basic needs seeking, thus, a harmony of coexistence between people and the environment that houses them.

The three main principles that best explain these issues will be listed (DIXON; SPOTTEN, 2014; WITHONEPLANET, 2017; HARLAND, 2018):

6. Caring for the future: Rethinking consumption

Issues that rethink consumption, related to the resources still available, must also be reconciled with social problems such as hunger. It is essential to see that it is possible to redistribute and make food accessible to different sections of society.

These issues of rethinking consumerism can be illustrated in relation to the natural resources available (Figure 7).

Figure 7. Mafalda and the powerful critique of values.

Source: Deposito de tirinhas, 2020.

7. Permaculture and modern coffee farming

The application of permaculture principles is of paramount importance for the implementation of organic and agroecological coffee farming. Conventional coffee is among the most chemically "treated" foods in the world, due to the large volumes of synthetic fertilizers, herbicides, fungicides, acaricides and insecticides used in farming.

Farmers have high exposure to agrochemicals when spraying crops and handling them during harvest. In the case of organic coffee, there is no use of synthetic fertilizers or chemicals used in cultivation or production, which means cleaner grains, air, land and water: coffee is grown only with organic fertilizers, such as coffee pulp, chicken manure or compost, and does not promote deforestation.

Indeed, it helps in the conservation of wildlife, birds, vegetation and the prevention of soil erosion and damage. Natural vegetation retains and increases the natural fertilisation of the soil. Organic or agro-ecological coffee farming also contributes to the conservation of wildlife, birds, vegetation and the prevention of erosion.

Natural vegetation promotes natural fertilisation of the soil via deposition of organic residues and their decomposition. Soil nutrients are retained and replenished naturally for sustainable agriculture that will continue to produce year after year (Figure 8).

Figure 8. Organic and agro-ecological coffee cultivation on Incaper's farm, Pacotuba, district of Cachoeiro do Itapemirim, ES (diversified system)

Source: Personal archive.

Table 2 compares coffee production methods with those used in other countries.

in organic and conventional systems.

Table 2 - Comparison of coffee production: conventional *versus* organic

	Conventional Agriculture	**Organic Agriculture**
General Objectives	Generally serving short-term economic interests.	Meeting economic interests: above all, self-sustaining ecological and social interests.
System Structure	Monoculture	Diversified system
How to Face the Ground	Like a physical substrate, a plant support.	As a living being (eminently biological environment)
Genetic Resources	Reduction of variability; susceptibility to the environment; transgenic species.	Environmental adaptation; resistance to the environment.
Fertilisation	Highly soluble fertilizers; unbalanced fertilization.	Recycling; ground rocks; organic materials; balanced and adequate nutrition.
How to deal with pests and diseases	Agrotoxics	Diversification and consortia; alternative controls.
System inputs	High capital and energy: little work.	Little capital and energy: A lot of work.
System outputs and consequences	Unbalanced and contaminated food; low product value; environmental aggression.	High biological value food; ecological balance; high product value; sustainability of the system.

Source: CPT, 2013.

On the other hand, Figure 9 shows a coffee plantation under conventional cultivation, single, without soil or plant protection (full sun).

Figure 9. Conilon coffee cultivation under conventional system at Incaper's farm, Pacotuba, district of Cachoeiro do Itapemirim, ES

Source: Personal archive.

8. Coffee farming and permaculture: how to set up a coffee plantation

The production of organic coffee has demonstrated a great opportunity for small producers to act responsibly and with products valued by the final consumer. When opting for the conversion to organic agriculture, the coffee grower must consider the need to adapt to the norms and regulations that will directly affect his current practice, with regard to production, labour and commercialisation process.

The Empresa Brasileira de Pesquisa Agropecuaria (Embrapa, 2015) emphasizes that in accordance with the norms of the International Federation of Organic Agriculture Movements (IFOAM), the conversion should follow an annual plan. The interested party should elaborate a conversion project, which should be previously presented to the certifying body, or the inspector, on the occasion of the first visit.

The characterization of the unit as organic will depend on the fulfilment of this plan. A contract must be signed between the farmer or

producer organisation and the certification body. The documentation of the rural establishment (general data, maps, history of planting areas) must be made available to the inspectors. Cash books must contain records of yields and product flow, including the stages of processing, storage, packaging and sale. A detailed list of inputs used is also required for approval. At the beginning of the conversation, social aspects, such as housing conditions, food and hygiene, will be inventoried and an improvement plan, if applicable, must be submitted (SEBRAE, 2020).

According to the same author, in the implementation of this plan, a timetable of execution shall be observed. Samples (soil, water, plants, harvested products, among others) may be taken by the certification body at any time for residue analysis. The transition shall correspond to the time that has elapsed since the date of the last application of non-permitted inputs in an agricultural area until it receives the organic seal. This period will depend on the extension of the productive unit, the environmental conditions, especially of the soil, and the technological level adopted by the coffee grower.

In units where crops are managed with minimal use of external inputs, eighteen (18) months will be sufficient to meet the requirements. On the other hand, highly technified or semi-technified production units will need a minimum period of three (3) years for the transition, time foreseen for the agrotoxics residues to be degraded in the soil (SEBRAE, 2020).

Conversion should be done in stages, replacing chemical fertilizers by organic ones. It is advisable to divide the production unit into uniform fields in terms of environment (soil, topography, sun exposure, among others). Based on these observations, the coffee grower should work towards converting 20 to 25% of the total area each year.

The use of agrotoxics should be stopped immediately, replacing them with preventive foliar sprays, using permitted grouts (Bordeaux, sulfocalcic, etc.) and biofertilizers, respecting the limit of use of these products (number of treatments and concentrations) and observing care in their handling. Ideally, the coffee grower should follow a schedule drawn up with an extensionist (ZACARIAS; SOUZA, 2019; SEBRAE, 2020).

CHAPTER IV

AGROFOREST SYSTEMS (SAFs) AND COFFEE FARMING

Joao Marcos Verly De Oliveira Da Silva [10]Mauricio Novaes Souza [11]Otacllio Jose Passos Rangel5 Mauricio Lorengao Fornazier [12]Geisa Correa Louback6 Grazielli Pirovani6 Camila Barbiero

[10] Bachelor in Coffee Technology. Ifes - Campus de Alegre, Caixa Postal 47, CEP: 29500-000, Alegre-ES. jmverly@gmail.com

[11] Dr. Professor of the Instituto Federal do Espirito Santo - Campus de Alegre, Caixa Postal 47, CEP: 29500000, Alegre-ES. mauricios.novaes@ifes.edu.br; ojprangel@ifes.edu.br

[12] Master's student in Agroecology in the Postgraduate Programme in Agroecology of Ifes - Campus de Alegre, Caixa Postal 47, CEP: 29500-000, Alegre-ES. fgzampieri@hotmail.com; mauzier if@hotmail.com

Siqueira [13]

1. Introduction

Agroforestry systems (SAFs) have been widely disseminated as agricultural exploitation models that greatly contribute to the sustainability of current agricultural exploitation.

In order for agricultural models to be classified as such they must follow the definition of the SAFs, in which it is necessary to use arboreal, shrub and herbaceous plants, associated with agricultural and forage species with or without the presence of animals, but obligatorily associated with forest species.

Such models become interesting alternatives for small farmers who seek to obtain an economically viable intensive exploitation. The term SAF corresponds to a form of land use and natural resource management in which woody species (trees, shrubs and palms).

2. Classification of SAFs

The classification of the SAFs is based on the criteria of spatial and temporal arrangement, the importance and role of the components, production planning or production of the system and its socio-economic characteristics (Nair, 1985 *apud* Santos 2000). SAFs, according to Souza (2004) and Bernardes (2008), can be classified according to their components into:

> **Silviagricola or agroforestry:** Forestry species and agricultural crops (Figure 1).

Figure 1. silviagricola or agroforestry model

[13] Student of Agroecology in the Post-Graduation Program in Agroecology of Ifes - Campus de Alegre, Caixa Postal 47, CEP: 29500-000, Alegre-ES.

Source: ABDO, 2008.

According to the arrangement of the species in the field, the models may have a great variation, consisting of mixed dense systems, such as home gardens, mixed low density systems, such as agro-forestry-pastoral systems, in strips or continuous or random. According to the arrangement of the species in time, the SAFs can be simultaneous or sequential (SOUZA, 2004).

According to Martins and Souza (2013), the sequential SAFs occur so that there is a time interval between the harvest of the first crop and the sowing of the subsequent crop. For the simultaneous ones, it can be observed that there are several situations: two crops with the same planting and harvesting epoch (coincident SAF), crops with the same sowing epoch and different harvesting epochs (concomitant).

> **Silvipastoral:** Forest and forage species for animal feeding (Figure 2):

Figure 2: Silvipastoral model

Source: Martins; Souza, 2013.

Agrosilvopastoral: Forest species, agricultural crops and forage for animal feed (Figure 3):

Figure 3. agro-forestry-pastoral model

Source: BERNARDES, 2008.

An interesting example of SAFs with concomitant cultures is the planting of the royal palm and the Jussara palm heart to obtain palm hearts under eucalyptus plantations, where the palm heart culture is extracted at five (5) years, therefore, before the end of the first eucalyptus cycle, which occurs at seven (7) years (Figure 4).

Another SAF model is the overlapping one, when a crop is sown before the end of the cycle of a crop already installed on the site, and the harvest will be done after the end of the cycle of the first crop installed.

Figure 4. royal palm under eucalyptus plantation

Source: BERNARDES, 2008.

There is also the interpolated model, in which during the cycle of a perennial crop, shorter cycle crops are planted, such as annual crops under rubber trees or eucalyptus.

3. Coffee in SAFs

Among about 100 *Coffea* species described (FAZUOLI, 1986; FERRAO et al., 2017), *Coffea arabica* L. (arabica coffee) and *Coffea canephora* Pierre (robusta coffee) are the only ones with economic expression in the world market.

Native to tropical regions of Africa, both arabica and robusta coffee evolved as woody understory species. The first arabica coffee plantations were therefore managed under shading, through intercropping with larger trees, to simulate the crop's natural habitat. In many situations, however, full sun coffee plantations can produce more than shaded ones (FOURNIER, 1988; BEER et al., 1998; FERRAO et al., 2017).

As a consequence, shading was abandoned as a regular cultural practice in many regions of the world. In Brazil, for example, this occurred from the 1960s onwards.

The physiological characteristics of coffee allow it to be cultivated in systems shaded by trees (DaMatta, 2004). Coffee SAFs can contribute to the conservation of biodiversity and the provision of environmental services, both for local populations and for the global society (RICE, 2008; MARTINEZ et al., 2009).

<* **Among the possible environmental services are:**

J Reduction of soil and water losses by surface run-off (CRASWELL et al., 1998);
J Greater root depth resulting in more efficient use of rainwater (HARMAND et al., 2007; CANNAVO, 2011);
J Increase in biological diversity (CARDOSO et al., 2010);
J Strategic tool for microclimate management against the risks of global climate change (LIN, 2007; SILES et al., 2010); and
J *To* increase the efficiency of radiation conversion (PEZZOPANE, 2007).

With regard to the profitability of the producer, it should also be considered that coffee is an agricultural product whose price is based on qualitative parameters and varies significantly depending on the quality presented. Therefore, adequate care and techniques in the management of the crop, as well as in the harvest and post-harvest phases, are fundamental for obtaining a quality product with better profitability (MALTA et al., 2012 *apud* SOUZA, 2018).

The fact is that the search for sustainability and quality, which are responsible for the opening of new consumer markets that increasingly demand coffee with higher quality parameters, has become essential for the permanence of man in the field, ensuring him and his family, better living conditions. For this it is necessary more studies directed towards a development that provides a low cost of production and that facilitates the obtainment of superior coffees (FERRAO et al., 2017).

3.1. Fruit species

Fruit trees or trees that can be used for charcoal production are viable alternatives to some leguminous trees commonly used to shade coffee plantations.

The implementation of SAFs requires careful planning and the long term must always be kept in mind. Although the ecophysiological interactions between the coffee plant and the other perennial species already manifest themselves in the short term (effects on the soil and on the microclimate), the economic results are usually observable in the long term. For this reason, the economic return of the system has to be calculated or estimated for the long term (DAMATTA et al., 2007).

In intercrops with fruit trees, such as coconut or banana, or palm heart producing species, such as pupunha, the economic analysis should be done from the beginning of the harvest of the fruit or palm heart, and equivalence indices can be established in relation to the monoculture of coffee or the intercrop (SALES; ARAUJO, 2005).

Particularly in Espirito Santo, growing conilon coffee with shade trees has become increasingly common. In a recent survey, Sales and Araujo (2005) found 27 intercropping plantations in 10 different municipalities, totaling about 115 ha (Table 1).

Of this total, the great majority is located in the northern region of the State, and the consortia with tree species such as Australian cedar, teak and rubber trees seem to be the most used. However, besides the tree species mentioned above, fruit species such as cashew, coconut and papaya also occupy an important place (Figure 5).

To implement an agroforestry system with the planting of native fruit trees in coffee plantations it is interesting to choose trees with the following characteristics (SOUZA, 2004): *J* Deep roots;

J Legume family;

J Plants resistant to pruning;

J Good biomass production; and

J Sexual reproduction is easily controlled.

Table 1. Survey of areas of conilon coffee intercropped with trees in the State of Espirito Santo

Source: DAMATTA et al., 2007.

Nome comum	Nome científico	Espaçamento[1] (m)	Municípios	Área total (ha)/ (nº de propriedades)
Cajueiro	Anacardium occidentale	10 x 12	Vila Pavão	5,0 - (1)
Coqueiro	Cocos nucifera	9 x 8	São Gabriel da Palha	4,5 - (1)
Cedro Australiano	Toona ciliate	3 x 3 e 15 x 9 m	Jerônimo Monteiro e Sooretama	31,0 - (2)
Grevilha	Grevillea robusta	3 x 6 m e diversos	Vila Pavão	0,2 - (1)
Ingá	Inga sp.	9 x 6 e 11 x 10 m	Iconha, São Domingos do Norte	1,2 - (2)
Nim Indiano	Azadirachta indica	6 x 6 m	Vila Valério	1,0 - (1)
Peroba	Paratecoma peroba	Diversos	Alegre	1,5 - (1)
Seringueira	Hevea brasiliensis	3 x 10 a 10 x 10 m (FS) 18 x 4 x 3 m (FD)	Vila Valério, São Gabriel da Palha	23,4 - (5)
Teca	Tectona grandis	8 x 8 m	Sooretama	30,0 - (1)
Urucum	Bixa orellana	6 x 3 m	São Gabriel da Palha	4,0 - (1)
Frutíferas, Madeiráveis e Seringueira	-	2 x 2 a 12 x 10 m, e diversos	São Gabriel da Palha, Nova Venécia, São Domingos do Norte, Rio Bananal	13,8 - (11)
Total	-	-	-	115,6 - (27)

Figure 5: Consortia of conilon coffee with some forest or fruit species: rubber tree in double row (A), teak at one year of age (B), coconut tree (C), Australian cedar (D), papaya and Australian cedar (E), implantation; F), formed consortium)

Source: DAMATTA et al., 2007.

The implementation of SAFs in organic coffee plantations offers advantages to the producer. According to Dubois (1996), establishment and maintenance costs can be kept within acceptable limits for the farmer. SAFs can increase family income.

The cultivation of fruit trees has a low maintenance cost and can generate a higher income than pasture or grazing land, besides helping to maintain or improve the productive capacity of the land, improving the physical structure of the land, as they accumulate a greater quantity of organic material, as well as providing greater protection of the environment, reducing the felling of trees and controlling erosion.

4.2. Tree species

The use of coffee in agroforestry systems can be an important alternative for small rural properties. The use of tree species in intercropping with coffee is commonly adopted in several countries, including Brazil.

The afforestation with species for multiple uses that add value to the coffee plantation becomes an interesting option as it can minimize climate change, act as windbreaks, shelter for natural enemies of pests and still represent an option of gain for the producer (SOUZA et al., 2020).

According to the same authors, the selection of the most appropriate forest species for intercropping and the interactions between crops are of fundamental importance for the improvement of the benefits resulting from the use of agroforestry systems in climatic and soil conditions.

The cultivation of tree species may mean additional income or a saving of values, which in a moment of crop renewal or of a climatic problem such as hail or drought can be used. On the other hand, the consumer market demands and pays a premium for products that come from more balanced production systems, with less use of pesticides and that are environmentally sustainable (SOUZA et al., 2020).

There are many tree species used in agroforestry systems in coffee plantations: Acacia *(Acacia mangium), Acrocarpus (Acrocarpus sp),* African Mahogany (*Khaya ivorensis),* Teak *(Tecton a grandis),* Macadamia *(Macadamia integrifolia)* (Figures 6 and 7):

Figure 6: Coffee plantation shaded with Teak, Cedar, Acacia and Avocado

Source: Assessoria de Comunicagao UFLA, 2016.

Figure 7. Planting of African mahogany intercropped with cocoa.

Source: SOUZA, 2018.

5. Final considerations

In Brazil there are already many successful consortia established in research institutes as well as in producer areas. However, it is necessary to evaluate quantitative and qualitative parameters of the biophysical variables of the socioeconomically interesting SAF's already existing in the rural environment.

The research shows that properly conducted SAFs are important tools in the context of sustainable development. They also constitute viable approaches for the recovery of degraded areas, including those degraded by coffee plantations.

From the socio-economic point of view, several authors point out advantages, such as: (a) increased income opportunities per unit of area; (b) greater variety of products and, or, services; (c) improved food and human nutrition, especially in the context of family farming; (d) crop diversity and risk reduction; (e) amortization of planting and forest management costs; (f) improved distribution of income and rural labour; (g) reduction of some cultural practices of the conventional system; and (h) contribution to landscape management.

However, for many authors, it is necessary to study more deeply several aspects related to agroforestry systems, such as nutrient cycling, hydrological cycle, economic analysis, potential species, animal component, erosion and pest and disease incidence, so that sustainability indicators can be better defined for these systems.

Although the SAF's are advocated as an alternative capable of promoting environmental and social changes, especially in humid tropical regions, socio-economic, cultural and political factors have prevented the creation of a scenario sufficiently attractive for the different segments of society to adopt this type of land use.

In the technological area, the non-adoption by small farmers is mainly centred on the lack of information on how to manage such complex and specific systems for each region. All these factors make it difficult to generalise the conclusions of research and extensionist recommendations. There is a need for more studies on the self-ecology of the species used.

Changes in the subsistence economy of the tropical regions, for a market economy of the SAF's, demand study, development and improvement of technologies for this modality of land use. It is hoped that these technologies will be capable of integrating anthropic and environmental concerns, avoiding both land degradation and unregulated exploitation of the forest, and improving the income of small producers, thus reducing rural exodus.

There is a need to develop studies on: a) the existing types of agroforestry systems and their components; b) the relationship between diversity and stability of the systems including the various indicator parameters (e.g. net yield and nutrient cycling); and c) local knowledge on the establishment, management and use of these systems and their components.

The studies should make small farmers aware of the need to exercise a critical sense about the markets and marketing of the products generated by the SAFs. Such research can make SAFs competitive with conventional farming and forestry alternatives.

As an agroecological practice that seeks to be sustainable, the SAFs have drawn attention in recent years by presenting a range of advantageous aspects to conventional models. However, as has been verified, there are many factors, biotic, abiotic and social, that are involved in an agroforestry system.

Many of the possible interactions that can be found in an SAF have not yet been studied. Therefore, the potential of this system of combining ecological aspects of natural ecosystems with agriculture and forestry stimulates further studies on SAFs. In view of the great acceptance and adoption by farmers in various regions, it is necessary to work on the dissemination of the systems already defined, with the participation of the producers in the whole process. There should be a systematic approach to rescue the importance of the trees in the production system, particularly through the deposit of litter and organic material in the soil, fundamental to the environmental recuperation procedures.

Considering, historically, that Brazilian riches, such as forests, have been exploited only extractively, they generate foreign exchange and development for other countries. For this reason, the country is in the imminent possibility of importing wood due to the possible "forest blackout". The SAF's can collaborate to change this contradiction, generating wealth from this asset, investing in technology and productivity, generating competitive differentials that add value to the material.

Incentive actions for planting and sustainable timber production are indispensable; however, they must be accompanied by stimulation of value addition, which generates employment and social inclusion. SAF's favor rational forest exploitation.

CHAPTER V

AGRO-ECOLOGICAL COFFEE-GROWING: SUGGESTIONS FOR RESEARCH

Joao Marcos Verly De Oliveira Da Silva [14]Mauricio Novaes Souza [15]Otacllio Jose Passos Rangel6 Jose Francisco Lopes6 Mauricio Lorengao Fornazier [16]

1. Conventional coffee growing in Brazil

In Brazil, coffee farming has promoted development and significant economic gain. It has achieved relevance throughout its history and allowed the country to excel

[14] Bachelor in Coffee Technology. Ifes - Campus de Alegre, Caixa Postal 47, CEP: 29500-000, Alegre-ES. jmverly@gmail.com
[15] Dr. Professor of the Instituto Federal do Espirito Santo - Campus de Alegre, Caixa Postal 47, CEP: 29500000, Alegre-ES. mauricios.novaes@ifes.edu.br; ojprangel@ifes.edu.br
[16] Master's student in Agroecology in the Post-Graduation Programme in Agroecology of Ifes - Campus de Alegre, Caixa Postal 47, CEP: 29500-000, Alegre-ES. fgzampieri@hotmail.com; mauzier_if@hotmail.com

in coffee production on the world stage. However, it was observed that the coffee production growth caused many negative socio-environmental impacts, clearly evidenced throughout its growth and development.

Several impacts were observed, such as (SOUZA, 2018):

J The introduction of coffee monocultures has led to increased deforestation of the Atlantic Rainforest and the Cerrado;
J The indiscriminate use of agrochemicals by coffee farmers, leading to poisoning and deaths of many farmers;
J The felling of riparian forests causing a degradation of water resources and their contamination; and
J The indiscriminate use of agrotoxics.

The fact is that this model has reduced the resistance and resilience of agroecosystems, generating environmental imbalance through monocultures, causing soil impoverishment and degradation. The reduction of biological resistance of the ecosystem favoured the emergence of pests and diseases with significant loss of productivity and production (SOUZA, 2018; SOUZA et al., 2020).

Monocultures in full sunlight have been the model adopted for coffee production in Brazil since the beginning of the 19th century. This form of plantation management causes a low level of biological diversity. The idea that coffee can be grown under the forest canopy is unknown to most Brazilian producers; however, it is already successfully adopted in Colombia, Venezuela, Costa Rica, Mexico, Nicaragua and Panama (AGUIAR- MENEZES et al., 2007).

Coffee production in conventional and agroecological systems differ mainly in the inputs used during coffee growing. The inputs used in the control of agricultural pests is still the main strategy, seeking greater productivity with lower costs. The agrochemicals used in the control of agricultural pests are of great concern because they are highly toxic and, therefore, their consumption and residual power in food, water and environment need to be controlled (CALDAS; SOUZA, 2000).

According to Lima et al. (2002) and Souza (2021), the soil conservation practices should be highlighted; the control of pests (insects) should have an appropriate ecological management; the ecological management of weeds and invasive plants with more efficient techniques; the correct disposal of effluents resulting from the pulping of coffee! These are essential conservationist practices which help to reduce the negative environmental impacts generated by agricultural activities.

2. Organic and agro-ecological coffee farming

Organic agriculture is a production model that receives this name because it does not use synthetic or agrotoxic fertilizers. This model suggests the use of green manure, crop rotation to avoid soil erosion, composting, biological pest control, among several other environmentally correct practices.

Organic coffee farming is conducted without the use of agrotoxics and synthetic fertilizers of high solubility, substituting them by a form of alternative natural use and natural sources such as the composting of organic material of vegetable and animal origin, bio-fertilizers, residue from coffee pulp and husk, vermicompost, among others (THEODORO, 2003).

According to SEBRAE (2018), organic production does not use agrotoxics and soluble fertilisers, genetically modified organisms and ionising radiation at any stage of the production process, processing, storage, distribution and commercialisation.

Organic farming has the following objectives:

J Offer healthy products free of intentional contaminants;

J Preserving the biological diversity of natural ecosystems and restoring or increasing the biological diversity of modified ecosystems where the production system is located;

J Increasing soil biological activity;

J Promote the healthy use of soil, water and air;

J Reduce to a minimum all forms of contamination of these elements which could result from agricultural practices;

J Maintain or enhance soil fertility in the long term;

J Recycle organic waste; and

J Reduce the use of non-renewable resources to a minimum.

This form of production is based on:

J In the use of renewable resources and locally organised agricultural systems;

J In encouraging integration between the different segments of the production chain;

J On the consumption of organic products;

J In the regionalisation of the production and trade of these products;

J In the handling of agricultural products based on the use of careful elaboration methods, with the purpose of maintaining the organic integrity and the vital qualities of the product at all stages.

Some suggestions for research in organic farming are listed below:

2.1. green manures

The use of green manures promotes nutrient cycling and biological nitrogen fixation, reduces erosion, improves soil structure, incorporates organic matter, increases the water retention capacity of the soil and reduces the incidence of weeds and pests. It has emerged as an option: it is an ancient practice used by many farmers nowadays with the aim of improving the physical, chemical and biological properties of soils, productivity and quality of crops of economic interest.

It also works in the management of spontaneous plants, in the adaptation of the property for the implantation of the agro-ecological and/or organic production system and, mainly, with the objective of reducing production costs.

One of its main and most important benefits is related to the use of plant species, mainly of the Fabaceae or Leguminosae family, which associate with nitrogen-fixing bacteria from the air, removing nitrogen (N) from the air and making it available in the soil for the plants. Soil fertility and inadequate use of liming and fertilizers, especially with N, are considered the main factors responsible for low productivity in areas intended for production (SOUSA et al., 2018; MARTINS et al., 2019).

According to the same authors, N is an essential macronutrient for full plant development and one of its greatest limiting factors. Nitrogen fertilization and the decomposition of organic matter control the availability of N in the soil for the plants, and when crops with a low ratio between carbon and nitrogen in dry matter are used in rotation, decomposition and mineralization are faster and N cycling occurs in a short period of time, reducing, however, the protection of the soil. On the other hand, the higher the C/N ratio of the residues, the slower the decomposition will be.

In this way, the use of green manure is a sustainable alternative in production models, also in coffee cultivation. The benefits of cover crops were observed by several authors, showing improvements in the chemical and physical properties of the soil (EIRAS; COELHO, 2011; SOUSA et al., 2018; MARTINS et al., 2019).

Among the effects of green manure on soil fertility, the following are highlighted: the supply of organic matter; greater availability of nutrients; and greater effective cation exchange capacity of the soil. In addition, these changes are important in the natural organic cycle, regulating biogeochemical processes and facilitating symbiotic relationships (CARDOSO et al., 2014; SOUSA et al., 2018).

Among the plants used in green manure, some legumes stand out: crotalaria *(Crotalaria juncea),* lab-lab *(Dolichos lablab),* mucuna preta *(Mucuna aferrima)* and pueraria *(Pueraria phaseoloides).* These species have the ability to biologically fix nitrogen and make it available to the succeeding crop (TEODORO et al., 2011; EIRAS; COELHO, 2011; FERRARI NETO et al., 2012).

Both tree, shrub and herbaceous species have a high degree of hardiness, high biomass accumulation, and also act as regulators of soil temperature and humidity and protect against the risk of erosion.

Furthermore, they increase the efficiency in nutrient cycling, mainly by the high C/N ratio, favoring agricultural and forestry crops (BERNARDES et al., 2010; LEITE et al., 2010). For this, it is necessary to know the peculiarities of each plant, in addition to its responses to the different conditions and consortia in which it will be planted (CARVALHO et al., 2015; TORRES et al., 2015; MIRANDA et al., 2018).

The analysis of nutrient cycling appears as a tool to know the dynamics of these relationships, indicating the power of establishment and ecosystem support of each individual. It is also possible to evaluate the quality and quantity of the maintenance service, nutrient addition and carbon stock in the soil (VITAL et al., 2004).

Various studies have shown that the green manuring technique is responsible for improving the physical, chemical and biological conditions of the soil. Legumes that use the association of nitrogen-fixing bacteria to transfer nitrogen to the plant and stimulate the development of the mycotic fungus population: these increase the volume of roots, favouring the absorption of water and nutrients by the roots. Nitrogen fixation can also be obtained by inoculation of the seeds; however, it is a more expensive technique (EMBRAPA, 2011; ZACARIAS et al., 2019).

Besides the aforementioned characteristics, legumes are chosen because they are rustic, have high production of dry matter and thanks to their root system, generally deep and branched, are able to extract nutrients from deeper soil layers and, when decomposed, return these nutrients to the soil (nutrient cycling), being important sources of organic matter of excellent quality (ARF et al. 1999; SOUZA, 2018).

In this way, the choice of the green manure species to be planted becomes very important, because each one has different characteristics, such as winter/summer planting, climate, contribution of green mass and dry mass, contribution of nitrogen and management such as pruning. The management of green manures is also very important, as the crotalaria, the feijao-de-porco and the feijao guandu, are some of the species used in intercropping, or even before planting coffee, according to (Figures 1 and 2).

Figures 1 and 2: Leguminosae in a coffee plantation

Source: Personal archive.

2.2. Ecological pest management (EPM) and integrated pest management (IPM)

The concern of the society for the consumption of food free of agrotoxics or chemical products has led the organic agriculture to adopt the ecological management of pests (EPM). This management has the purpose of promoting the management of the agroecosystems with the function of recreating a harmonic and natural environment, incorporating the basic ecological information to the integrated pest management (IPM).

Figure 3 represents a model of ecological weed control using cover crops that provide weed suppression through smothering processes in coffee tillage.

Integrated management can be used to manage insects, weeds and pests. An example of insect control is the use of microbiological insecticides, in which the *Beauveria bassiana* fungus colonises adult coffee berry borers.

Figure 3: Coffee plantation with weed management and use of leguminous plants between the rows

Source: Personal archive.

IPM allows us to know the habits of the pests existing in the field, enabling us to choose the best control methods to be used.

According to Theodoro (2006):

> "In fact, IPM has been used without full knowledge of the ecological interactions involved in the agro-ecosystem and, thus, therapeutic control measures are used, without really knowing the reasons that led certain insects to reach the *status of pest and how the* limiting *agents of the* population growth of these insects *act*" (Venzon et al., 2001, *apud* THEODORO, 2006).

The fight against insects, invasive plants and diseases in organic coffee farming is based on an ecologic way without damaging the environment, such as crop rotation, green manure, biological control of pests and diseases. It seeks to maintain the structure and productivity of the soil, working in harmony with nature.

It is through experience with conservationist agricultural practices that one learns the importance of organic farming, whose main role is to provide development and a new way of relating to nature and the environment.

2.3. Management between the lines

The maintenance of natural vegetation between the rows by means of mechanical weeding can reduce the production of fruits, principally in regions with high water deficits. The continued use of this practice favours an increase in the incidence of grasses, such as gingergrass *(Paspalum maritimum* Trind) to the detriment of other species, thus contributing to an increase in competition for soil moisture and nutrients, especially nitrogen (MESQUITA et al., 2016).

According to these same authors, on the other hand, the elimination or reduction of natural vegetation between the rows has a positive effect on the increase in production. It is up to the producer, therefore, to evaluate which management system should be used in his property, and which best suits his production system, selecting sustainable management practices that allow him to conciliate production with other economic activities.

In the spaces available in the planting line of the tree species, the giant guandu can be planted, and managed (pruned) whenever necessary, to avoid excessive shading. The straw of the giant guandu can be distributed in the line of trees, which besides providing nutrients, avoids and keeps weeds under control, avoiding weeding and frequent weeding (SOUZA et al., 2020).

As green manure between the rows, besides the giant guandu, one can also use crotalaria, mucuna preta, feijao de porco and also sorghum, always in a rotation system, at least until the first three years of planting the crop. The weeds present in the coffee crop, contrary to what was thought in the past, are very favorable for the coffee agro-ecosystem environment.

According to Ventura et al. (2007) and Zacarias et al. (2019), it contributes to:

JAumento of humidity;

Decrease in temperature;

JFavoring the development of soil microfauna;

JPromotes good conditions for fungal pathogens parasitoids of coffee pests;

J Helps increase plant material, such as organic matter content and helps nutrient cycling by making nutrients accessible to the plant;

Above all, it contributes to the attraction of biological control insects, mainly of the order Hymenoptera, which includes wasps and marimbondos, which promote the control of the leaf miner bug and acaros, among others.

Thus, the "weed" became an important ally in the conservation of soil and water and in soil fertility, improving the productivity of coffee plantations: the term weed control is no longer used among farmers. Some said that the ideal would be to keep the coffee plantations "clean", so that there would be no competition of nutrients, with this there was constant use of hoes, harrows, rotary tillers, leaving the soil exposed to the sun and to erosion (Picture 4).

Figure 4: Newly planted coffee plantation with brachiaria in the streets

Source: Personal archive.

Currently, "Bush Control" is being replaced by the "Bush Management" technique, which focuses on two pillars: keeping the coffee row without live weeds, but covered by straw; and the inter-row always with bush - the more bush produced, the better. The idea is to produce weeds in the inter-row and place them in the coffee row to form a "cushion" of dead plant material (Figure 5).

Figure 5. bush between the rows of coffee trees on the APRomero System farm in Sao Roque de Minas

Source: Personal archive.

With this, the coffee row remains always covered by straw and the inter-row by weeds growing and being continuously grazed, replenishing nutrients previously lost by surface erosion in the exposed soil: they are now taken up by the weeds and returned to the coffee in the form of organic matter (CafePoint, 2018).

This management between the rows was forbidden some years ago, because it was known as a competitor for nutrients and water with the coffee plant. On the other hand, new techniques have been improved and the use of weeds in the coffee rows has become necessary to improve the production and sustainability of the coffee growing environment.

According to Zacarias et al. (2019), the bush helps in the vegetation cover and the decomposition of the organic matter is made available to the plants in the form of nutrient by the cycling process, carried out by microorganisms in the soil (Figure 6).

Weed management between coffee rows on a farm belonging to Group A P Agricola (APRomero System), in Sao Roque de Minas

Source: Personal archive.

2.4. Substrates

Substrate preparation is very important for obtaining quality seedlings. In the case of scarcity of natural resources, there is a growing demand for alternative materials to be used for growing seedlings. These materials should be easily obtainable, environmentally friendly, have a stable structure, a reasonable decomposition time, be homogeneous, low cost and contain physical, chemical and biological characteristics compatible with the seedling to be produced (Figure 7).

For this, in the proposal of new materials, it is necessary to carry out studies concerning the physical/chemical quality of the same, as well as its adaptation and development of the plants. Finally, the challenge of searching alternative substrates in the different regions of the country, taking into account the waste from different activities/cultures, as well as the cost/benefit relation and the sustainability of the systems, is evident.

Gomes and Silva (2004) report:

> "Substrate is the medium in which the roots proliferate to provide structural support to the aerial part of the seedling and also to meet the water, oxygen and nutrient requirements. When choosing a substrate as a growing medium for seedlings, some physical and chemical characteristics related to the species to be planted must be considered, besides economical aspects. Such characteristics are homogeneity, low density, high porosity, good water retention capacity, high cation exchange capacity, good aggregation of particles at the roots, freedom from pathogens and undesirable seeds, easy handling at any time, abundance and economic viability".

Figure 7. Conventional substrate: "Seu Didi" nursery in Celina, Alegre, ES

Source: Personal archive.

Gonpalves and Poggiani (1996) report that while the seedlings remain in the nursery, the factor that must be taken into consideration is the substrate that will be used in the production of seedlings. It should possess characteristics such as consistency, good structure, high water retention capacity and high porosity.

The substrate should not expand, contract or present toxic substances and should be available and standardized. Pedroza et al. (2003) report that studies related to the components of the substrate are of great importance and that the promotion of organic and industrial residues is necessary to obtain seedlings with greater quality and at a lower cost.

Thus, sewage sludge appears as a viable alternative to be used as a component of substrate for seedling production. Guerrini and Trigueiro (2004) state that such procedures improve the physical, chemical and biological properties of the substrate, resulting in better quality seedlings to be taken to the field.

The residues generated from the processing of rice are also a much used substrate. Trigueiro and Guerrini (2003) point out that in recent years, rice husk has been intensively used as a substrate for plant growth because it is easily found and has characteristics favourable to plant development.

Soares (2014) states that this type of material presents high water retention capacity, fast and efficient drainage, providing good oxygenation to the roots, high aeration space to the substrate, resistance to decomposition, relative stability of structure, low density and pH close to neutrality.

Another organic residue with potential utilization is the coffee husk produced during its processing. This waste, being of organic nature, can be used in agriculture as fertilizer in crops and by nurserymen in the production of seedlings.

2.5. Organic waste and fertilisers

Organic fertilizers are also an important ally in organic agriculture. In coffee growing, the application of organic residues has a significant importance because it promotes the supply of nutrients and improves the physical, chemical and biological properties of the soil.

According to Lopes et al. (1975) and Souza (2015), Brazilian soils generally present low natural fertility and the organic matter contents are always low. The use of organic residues can contribute in a significant way to improve this condition, with low cost for the producer. For Rena et al. (1986), the organic fertilizers supply nutrients for the plants, which are liberated in the mineralization process.

2.6. Alternative Defenders

In the case of protection, alternative pesticides may be used. Theodoro et al. (2001) reiterate that in organic agriculture phytosanitary control is basically done by preventive measures, such as planting with more resistant varieties, organic fertilizer, mulching, intercropping and selective management of weeds and the use of windbreaks. The combination of these techniques, besides generating stronger plants, promotes an ecologically balanced environment, influencing the biological control of pests and diseases.

The Normative Instruction of the Ministry of Agriculture and Supply - MMA (1999) provides, in its annex III, about the means against fungal diseases and means of protection in the production of organic vegetables, recommending the adoption of some management practices: Simple sulphur and its preparations, only as fungicides; plant extracts; compound extracts; vermicomposts; Bordeaux and sulphocalcic solutions, at the certifier's discretion; homeopathy; preparations that stimulate plant resistance, with an inhibiting action against certain pests and diseases such as medicinal plants, propolis, lime and seaweed extracts, rock powder and similar.

Thus, Pedin (2000) points out that when the product used does not harm either man or the environment, it can be used to grow organic coffee.

◆ "Neem" oil

An organic insecticide that has been studied is "neem" oil. This oil is taken from the fruit of the neem tree, which consists of an almond found inside the seed produced by the neem tree.

All these almonds are cleaned and the shell removed, then crushed and cold-pressed to obtain neem oil. The cake resulting from the pressing of the seeds can be incorporated into the soil in order to control various types of fungi. The oil obtained is rich in fatty acids, mainly oleic acid, linoleic, palmitic, stearic and arachidonic, neem oil has many properties.

Neem oil is well known as an organic insecticide that can be applied to plants to combat pests and pathogenic fungi. Neem oil has become advantageous for the home and farm grower because it contains several fungi, acts as a bactericidal insecticide and can also control nematodes when applied in the form of neem cake associated with castor bean cake.

Its use can be combined with the practices of biological pest control, since it does not affect mammals or the environment. Moreover, neem oil is a cheap and viable option for both the farmer and the home grower.

FINAL CONSIDERATIONS

Generally speaking, it is possible to say that the conventional agricultural management system used by today's coffee farming is characterised by the artificialisation and simplification of agro-ecosystems, depending exclusively on inputs that are external to the system, such as agrotoxics, acaricides, fungicides, insecticides and soluble fertilisers - in addition to degradation, they increase production costs excessively and make the producer a hostage of this production model.

The devastation of extensive forest areas and the contamination of water and soil resources, associated with the constant use of inputs, causes constant imbalance to the environment. Such conditions tend to alter the processes of self-regulation of pests and diseases, reduce the resilience and recuperation power of the ecosystem against climatic and phytosanitary adversities, and deregulate the stability, flexibility, equity and self-sufficiency that diversified agroecosystems possess.

In opposition to this system, sustainable coffee farming models can contribute to the conservation of biodiversity and agro-biodiversity, of traditional and peasant culture, enabling the perpetuation of family coffee farming from an perspective that transcends coffee production and encompasses greater desires such as the social reproduction of families in rural areas, the quality of life of farmers and the preservation of natural resources for future generations. The expansion of local sustainable agriculture will be successful if it is always based on the concept of a harmonious relationship with nature.

The use of substrates, defensives and alternative fertilizers, as well as consortia (SAFs and Permaculture), among other means of research, will favour the development of a sustainable system for the management of coffee.

Sustainable rural development is not only about improving environmental quality, it is also about socio-economic, ethical and cultural issues.

Ecological coffee farming provides farmers with opportunities to reach different markets and add value to the coffee they sell, improving the quality of life of the families involved and guaranteeing the succession process for their descendants.

BIBLIOGRAPHICAL REFERENCES

AGUIAR-MENEZES, E. L. et al. Susceptibility of six Arabica coffee cultivars to fruit flies (Diptera: Tephritoidea) in an organic system with and without afforestation in Valenga, RJ. **Neotropical Entomology**, Londrina, v. 36, n. 2, p. 268273, 2007.

ANJOS, K. M. G. dos. **Investigacao e avaliagao da toxicidade aguda dos agrotoxicos mais utilizados no cinturao verde da Grande Natal (RN/ Brasil) para o peixe-zebra (Daniorerio Hamilton Buchanan, 1822, Teleostel, Cyprinidae)**. 2009. 63 f. Dissertation (Master in Aquatic Bioecology) - Universidade Federal do Rio Grande do Norte, Natal, 2009.

ARF, O. et al. Efeito da rotacao de culturas, adubagao verde e nitrogenada sobre o rendimento do feijao. **Revista Pesquisa Agropecuaria Brasileira**, Brasilia, v.34, n.11, p.2029-2036, 1999.

ARMANDO, M. S. **Agrofloresta para a agricultura familiar**. Brasilia: Embrapa Recursos Geneticos e Biotecnologia, 2002. 11p.

BARCHI, R. Educagao ambiental e (eco) governabilidade. Universidade de Sorocaba, Department of Geography, Sorocaba, SP, Brazil. **Ciencias e Educagao**, Bauru, v. 22, n. 3, p. 635-650, 2016.

BEER, J.; MUSCHLER, R.; KASS, D.; SOMARRIBA, E. Shade management in coffee and cacao plantations. **Agroforestry Systems**, v. 38, p. 139-164, 1998.

BERNARDES, M. S. **Sistemas agroflorestais**. In: XXXIII SECITAP. Jaboticabal: UNESP, Lecture. 2008.

BERNARDES, T. G.; SILVEIRA, P. M.; MESQUITA, M. A. M.; AGUIAR, R. A.; MESQUITA, G. M. Decomposition of biomass and release of nutrients from brachiaria and mombaga, under cerrado conditions. **Pesquisa Agropecuaria Tropical**, Goiania, v. 40, n. 3, p. 370-377, 2010.

BLACKWELL, R. D.; MINIARD, P. W.; ENGEL, J. F. **Consumer behavior.** Tradugao: Eduardo Teixeira Ayrosa. 9. ed. Sao Paulo: Thomson, 2011.

BRASIL. **Instrugao Normativa N.007 de 17 de mayo de 1999**. Estabelece normas para produgao de produtos organicos vegetais e animais. Diario Oficial da Uniao, Brasilia, n.94, Segao 1, p. 11, 19 May 1999.

BRASIL. **National curriculum parameters:** third and fourth cycle of basic education. Temas transversais. Brasilia: MEC/SEF, 1998.

CAFEPOINT. **Cup of Excellence champion seeks to encourage young coffee growers in the region**. Available at: https://www.cafepoint.com.br/noticias/giro-de-noticias/campeao-do-cup-of-excellence/. Accessed on: 11 Dec. 2018.

CALDAS, E. D.; SOUZA, L. C. K. R. Avaliagao de risco cronico na ingesto de residuos de pesticidas na dieta brasileira. **Revista de Saude Publica**, Sao Paulo, v. 34, n. 5, p. 529-537, 2000.

CANNAVO, P.; SANSOULET, J.; HARMAND, J. M.; SILES, P.; DREYER, E.; VAAST, P. Agroforestry associating coffee and *Inga densiflora* results in complementarity for water uptake and decreases deep drainage in Costa Rica. **Agriculture, Ecosystems&Environment, v.**140, n. 1-2, p.1-13, 2011.

CARVALHO, A. M.; COSER, T. R.; REIN, T. A.; DANTAS, R. A.; SILVA, R. R.; SOUZA, K. W. Manejo de plantas de cobertura na floraga e na maturate fisiologica e seu efeito na produtividade do milho. **Pesquisa Agropecuaria Brasileira**, v. 50, n. 7, p. 551-561,2015.

CARVALHO, I. C. M. **A invengao ecologica:** narratives and trajectories of environmental education in Brazil. 2. ed. Porto Alegre: Editora da UFRGS, 2002. 232 p.

CARVALHO, V. L. de; CHAULFOUN, S. M. **Doengas do cafeeiro:** diagnose e controle. Belo Horizonte: EPAMIG, 2000. 44p. (Boletim Tecnico, 58).

CARVALHO, Y. M. C. de. Agricultura organica e o comercio justo. **Cadernos de Ciencia & Tecnologia,** Brasilia, v. 19, n. 2, p. 205-234, 2002.

CORTEZ, F. **Educagao ambiental:** a formagao do sujeito ecologico. 4. ed. Sao Paulo: A, 2004b. 256 p.

CRASWELL, E. T.; SAJJAPONGSE, A.; HOWLETT, D. J. B.; DOWLING, A. J. Agroforestry in the management of sloping lands in Asia and the Pacific. **Agroforestry Systems**, v. 38, n. 1-3, p. 121-137. 1998.

DAMATTA, F. et al. **O cafe conilon em sistemas agroflorestais.** Cafe Conilon. Incaper, Vitoria, 2007.

DAMATTA, F. M. Ecophysiological constraints on the production of shaded and unshaded coffee: a review. **Field Crops Research**, v. 86, n. 2-3, p. 99-114, 2004.

DANTAS, G. C. da S. "**Educagao Ambiental"**; Brasil Escola. Available at: < http://brasilescola.uol.com.br/educacao/educacao-ambiental.htm>. Accessed on: 02 jun. 2020.

DEAN, W. **A ferro e fogo:** a historia e a devastação da Mata Atlantica brasileira. 3.ed. Sao Paulo: Companhia da Letras, 1997.

EIRAS, P. P.; COELHO, F. C. Utilizacao de leguminosas na adubagao verde para a cultura de milho. **Revista cientlfica internacional**, year 4, n. 17, p. 96-124, 2 011.

EMBRAPA. **Adubagao verde.** Seropedica. Embrapa Agrobiologia. 2011.

ESTEVES, J.; SILVESTRE, L. **Cafes sustentaveis:** Espirito Santo promoted Langamento do Curriculo de Sustentabilidade do Cafe em ambito nacional. Available at: < https://www.embrapa.br/busca-de-noticias/>. Accessed on: 10 Jul. 2020.

FAZUOLI, L. C. **Genetica e melhoramento do cafeeiro**. In: RENA, A. B.; MALAVOLTA, E.; ROCHA, M.; YAMADA, T. (Eds.). Cultura do Cafeeiro: fatores que afetam a produtividade. Piracicaba: Potafos, 1986, p. 87-114.

FERNANDES, J. M.; SIQUEIRA, L. C.; GARCIA, F. C. P. Agrobiodiversity in agroecological production systems. In: L. C. E. A. Ming (Ed.). **Agrobiodiversity in Brazil:** experience and paths of research. Recife: NUPEEA, 2010. Agrobiodiversity in agroecological production systems, p.75-94.

FERRAO, R. G.; FONSECA, A. F. A. da; FERRAO, M. A. G.; MUNER, L. H. de (Technical editors). **Cafe Conilon**. 2. ed. atual. e ampl. 2a reimpressao - Vitoria, ES: Incaper, 2017. 748 p.

FERRARI NETO, J.; CRUSCIOL, C. A. C.; SORATTO, R. P.; COSTA, C. H. M. Consorcio de guandu-anao com milheto: persistência e liberação de macronutrientes e silicic da fitomassa. **Bragantia, Campinas,** v. 71, n. 2, p. 264-272, 2 012.

FOURNIER, L. A. The cultivation of coffee (Coffea arabica L.) in sun or shade: an agronomic and ecophysiological approach. **Agronomia Costarricense**, v. 12, p. 131146, 1988.

FRAZER, P.; CHILVERS, C.; BERAL, B.; HILL, M. J. Nitrate and human cancer: a review of the evidence. **International Journal of Epidemiology**, v. 9, p. 311, 1980.

GARCIA, E. G. **Agrotoxicos e Prevenpao:** manual de treinamento. Sao Paulo:

Fundacentro, 1991.

GARCIA, E. G.; ALMEIDA, W. F. de. Exposure of rural workers to agrotoxics in Brazil. **Revista Brasileira de Saude Ocupacional**, v. 19, n. 72, p. 711, 1991.

GLIESSMAN, S. R. **Agroecologia:** processos ecologicos em agricultura sustentavel. 4. ed. Porto Alegre: Ed. Universidade/ UFRGS, 2009. 654p.

GOMES, J. M.; SILVA, A. R. da. Os substratos e sua influência na qualidade de mudas. In: BARBOSA, J. G. et al. (Ed.) **Nutripao e adubapao de plantas cultivadas em substrato**. Viposa: UFV, 2004. p. 190-225.

GONQALVES, J. L. M.; POGGIANI, F. Substratos para produgao de mudas florestais. In: SOLO 96 - SUELO CONGRESSO LATINOAMERICANO DE CIENCIA DO SOLO, 13, 1996, Aguas de Lindoia. **Expanded abstracts...** Aguas de Lindoia: SLCS/SBCS, 1996.

HARMAND, J. M.; AVILA, H.; DAMBRINE, E.; SKIBA, U.; E DE MIGUEL, S.; RENDEROS, R. V.; OLIVER, R.; JIMENEZ, F.; BEER, J. Nitrogen dynamics and soil nitrate retention in a Coffeaarabica-Eucalyptus deglupta agroforestry system in Southern Costa Rica. **Biogeochemistry**, v. 85, n. 2, p. 125-139, 2007.

HERZOG, F. Multipurpose shade trees in coffee and cocoa plantations in Cote d'Ivoire. **Agroforestry Systems**, v.27, p. 259-267, 1994.

LAYRARGUES, P. P. **Educapao Ambiental Crltica:** nome e enderepamentos da educapao. In: Identidades da educapao ambiental brasileira. Brasilia: Ministerio do Meio Ambiente, 2004a.

LEFF, E. **Saber ambiental:** sustentabilidade, racionalidade, complexidade, poder. Petropolis, RJ: Vozes, 2001.

Law, N°. 9.795, of 27 April 1999. **Dispoe sobre a educapao ambiental, institui a Politica Nacional de Educapao Ambiental e da outras providencias.** Brasilia, DF, 1999.

LEITE, L. F. C.; FREITAS, R. C. A.; SAGRILO, E.; GALVAO, S. R. S. Decomposipao e liberapao de nutrientes de residuos vegetais depositados sobre Latossolo Amarelo no cerrado maranhense. **Revista Ciencia Agronomica**, Fortaleza, v. 41, n. 1, p. 29-35, 2010.

LIMA, G. F. da C. Crise ambiental, educapao e cidadania: os desafios da sustentabilidade emancipatoria. In: LOUREIRO, C. F. B.; LAYRARGUES, P.; CASTRO R. S. de (Orgs.). **Educapao ambiental:** repensando o espapo da cidadania. 2a ed. Sao Paulo: Cortez, 2002.

LIN, B. B. Agroforestry management as an adaptive strategy against potential microclimate extremes in coffee agriculture. **Agricultural and Forest Meteorology,** v. 144, n. 1-2, p. 85-94, 2007.

LIN, B. B. Agroforestry management as an adaptive strategy against potential microclimate extremes in coffee agriculture. **Agricultural and Forest Meteorology**, v. 144, n. 1-2, p. 85-94, 2010.

LOPES, A. S. **A survey of the fertility of soils under "Cerrado" vegetation in Brasil** - Tese de Mestrado - North Carolina State Univ. 1975

MAIA, J. S. S. Educagao ambiental socio-historica como perspectiva para a reflexao-agao sobre o trabalho pedagogico nos primeiros anos da educação fundamental. In: TOZONI-REIS, M. F. C.; MAIA, J. S. S. (Coord.). **Educagao ambiental a varias maos:** educagao escolar, curriculo e politicas publicas. Araraquara: Junqueira e Marin, 2014. p. 26-40.

MAIA, J. S. S. **Educagao ambiental crltica e formagao de professores**. 1. ed. Curitiba: Appris, 2015. 241 p.

MALTA, M. R.; THEODORO, V. C. A. de; CHAGAS, S. J. R. Caracteristicas fisicoquimicas e sensorial de cafe beneficiado conduzido sob o sistema organico no municipio de Paraisopolis/MG. In: SIMPOSIO DE PESQUISAS DOS CAFES DO BRASIL; SIMPOSIO DE PESQUISA DOS CAFES DOBRASIL, 2003, Porto Seguro. **Annals.** Brasilia: EMBRAPA Cafe, 2003. p. 258-258.

MARTINEZ, M. L.; PEREZ-MAQUEO, O. V.; CASTILLO-CAMPOS, G.; GARCIA-FRANCO, G.; MEHLTRETER, J.; EQUIHUA, K.; LANDGRAVE, M.; ROSARIO, A. Effects of land use change on biodiversity and ecosystem services in tropical montane cloud forests of Mexico. **Forest Ecologyand Management**, v. 258, n. 9, p.1856-1863, 2009.

MARTINS, C. R.; GOMES, V. B.; WOLFF, L. F.; CARDOSO, J. H. **Leguminosae in fruitculture:** use and integragao in family properties in southern Brazil - Brasilia, DF: Embrapa, p. 66, 2019.

MARTINS, M. C.; SOUZA, M. N. Uma analise das variaveis do desenvolvimento rural sustentavel no uso da Integragao Lavoura Pecuaria e Floresta (ILPF) em municipios da Zona da Mata de Minas Gerais. Sustainable multifunctionality in the field. **Agricultura, pecuaria e florestas,** v.5, p.10-15, 2013.

MESQUITA, C. M. de et al. **Manual do cafe:** manejo de cafezais em produgao. Belo Horizonte: EMATER-MG, 2016. 72 p.

MESQUITA, C. M. de et al. **Manual do cafe:** implantagao de cafezais *Coffea arabica* L. Belo Horizonte: EMATER-MG, 2016. 50 p.

MIRANDA, S. C.; MATA, C. R.; FONSECA, K. S.; CARVALHO, P. S. Apontamentos sobre mudangas climaticas na agricultura Brasileira. **Revista Enciclopedia Biosfera**, v. 15, n. 27, p. 95-106, 2018.

MONTAGNINI, F. A. **Sistemas Agroflorestales:** principios y aplicacionesenlos tropicos. San Jose, Costa Rica: II CA. 622p.

MYERS, N.; MITTERMEIER, R. A.; MITTERMEIER, C. G.; DA FONSECA, G. A. B.; KENT, J. Biodiversity hotspots for conservation priorities. **Nature,** v. 403, n.6772, p. 853-858. 2000.

NAIR, P. K. R. 1989. **Agroforestry systems in the tropics.** Dordrecht: Kluwer Academic, 664p. (Forestry Sciences,31).

OCAMPO, J. A. **O mercado mundial de cafe e o surgimento da Colombia como um pals cafeicultor.** Rio de Janeiro, Revista Brasileira de Economia, vol. 37, n.4.

OLIVEIRA, T. K. de. **Caracterizapao de dois modelos de consorcios agroflorestais, Indices tecnicos e indicadores de viabilidade financeira/ Rio Branco**, AC: Embrapa Acre, 2010.

PEDROZA, J. P. et al. Produpao e componentes do algodoeiro herbaceo em funpao da aplicacao de biossolidos. **Revista Brasileira de Engenharia Agricola e do Ambiente**, v. 7, n. 3, .p. 483-488, 2003.

PEZZOPANE, J. R. M.; PEDRO JUNIOR, M. J.; GALLO, P. B. Balanpo de energia em cultivo de cafe a pleno sol e consorciado com banana "Prata Ana". **Rev. Bras. de Agron.**, v.15, n.2, p.169-177. 2007.

PRADO J. C. **Historia economica do Brasil.** 10. ed. Sao Paulo: Brasiliense, 1967.

REIS, P. R.; SOUZA, J. C.; VENZON, M. Manejo ecologico de pragas do cafeeiro. **Informe Agropecuario**, v. 23, p. 84-99, 2002.

RENA, A. B.; MALAVOLTA, E.; ROCHA, M.; UAMADA, J. **Cultura do cafeeiro:** fatores que afetam a produtividade. Potafos, 1986, 447p.

REYNOLDS, J. S. **Soil nitrogen dynamics in relation to groundwater contamination in the Valle Central, Costa Rica.** PhD Thesis, University of Michigan, MI, USA, 1991.

RICCI, M. dos S. F. et al. Growh rate and nutritional status of an organic coffee cropping system. **Sci. Agric.** Piracicaba, v.62, n.2, p.138-144, 2005.

RICCI, M. dos S. F. et al. **Influence of green manure on growth, nutritional status and productivity of coffee grown organically.** Seropedica, RJ: Embrapa Agrobiologia, 2002.

29p. (Documentos, 153).

RICE, R. A. Agricultural intensification within agroforestry: The case of coffee and wood products. **Agriculture, Ecosystems & Environment, v.** 128, n. 4, p. 212218. 2008.

RUSCHEINSKY, A. **Educapao ambiental:** abordagens multiplas. Porto Alegre: Artmed, 2002.

SANTOS, M. J. C. 2000. **Avaliapao economica de quatro modelos agroflorestais em areas degradadas por pastagens na Amazonia ocidental**. Piracicaba: ESALQ-USP, 75p. (Master's thesis).

SIAG - AGRO-METEOROLOGICAL INFORMATION SYSTEM. **Average data of the historical series of the meteorological station located in the municipality of Linhares-ES** 2006.

SILES, P.; VAAST, P.; DREYER, E.; HARMAND, J. Rainfall partitioning into throughfall, stemflow and interception loss in a coffee *(Coffea arabica* L.) monoculture compared to an agroforestry system with *Inga densiflora*. **Journal of Hydrology**, v.395, n.1-2, 2010/12/6/, p.39-48. 2010.

SOARES, W. L. **Uso dos agrotoxicos e seus impactos** a saúde **e ao ambiente:** uma avaliapao integrada entre a economia, a saúde publica, a ecologia e a agricultura. Available at: < http://bvssp.icict.fiocruz.br/pdf/ 25520_tese_wagner_25_ 03.pdf >. Accessed on: 17 abr. 2014.

SOUSA, I. R. L de; PAULETTO, D.; LOPES, L. S. de S.; RODE, R. Decomposition of species used as green manure in an experimental agroforestry system, Santarem, Para. **Agroecosystems**, v. 10, n. 2, p. 50 - 63, 2018.

SOUZA, I. I. de M.; ARAUJO, E. da S.; JAEGGI, M. E. P. C.; SIMAO, J. B. P.; ROUWS, J. R. C.; SOUZA, M. N. Effect of Afforestation of Arabica coffee on the physical and sensorial quality of the bean. **Journal of Experimental Agriculture International**, v. 42, n. 7, p. 133-143, 2020.

SOUZA, I. I. de M.; ARAUJO, E. da S.; JAEGGI, M. E. P. C.; SIMAO, J. B. P.; ROUWS, J. R. C.; SOUZA, M. N. Effect of Afforestation of Arabica Coffee on the Physical and Sensorial Quality of the Bean. **Journal of Experimental Agriculture International**, v. 42, n. 7, p. 133-143, 2020.

SOUZA, M. C. M. de. **Sustainable coffees and denomination of origin: the quality certification in the differentiation of organic, shade-grown and solidarity coffees.** 2006. 177f. Thesis (Doctorate in Environmental Science) - University of Sao Paulo, Sao Paulo, 2006.

SOUZA, M. N. **Anthropic degradation and environmental recovery procedures.** Balti, Moldova, Europe: New Academic Editions, 2018, v.1000. 376p.

SOUZA, M. N. **Degradapao e Recuperapao Ambiental e Desenvolvimento Sustentavel**. Viposa, MG: UFV, 2004. 371 p. Dissertação (Mestrado em Ciencia Florestal) - Universidade Federal de Viposa, 2004.

SOUZA, M. N. **Changes in land and water use and the management of natural resources.** Frankfurt, Germany: New Academic Editions, 2015, v.5000. 376 p.

SOUZA, M. N. **Topicos em recuperapao de areas degradadas.** VOL. I. CANOAS: Merida Publishers, 2021.133p.

SOUZA, T. S. **Qualidade da bebida do cafe conilon consorciado com diferentes especies arboreas e frutiferas, sob manejo organico.** Dissertation (Master's Degree) of Instituto Federal de Educacao Ciencia e Tecnologia do Espirito Santo Alegre Campus. *Stricto Sensu* post-graduation program in Agroecology. 2018. 57p.

TEODORO, R. B.; OLIVEIRA, F. L. de; SILVA, D. M. N. da; FAVERO, C.; QUARESMA, M. A. L. Aspectos agronomicos de leguminosas para adubapao verde no Cerrado no Alto Vale do Jequitinhonha. **Revista Brasileira de Ciencia do Solo**, v. 35, p. 635-643, 2011.

THEODORO, V. C. A. de et al. Alterapoes da qualidade de graos de cafes *(Coffea arabica* L.) colhidos no pano e no chao, provenientes de sistemas de manejo organico, em conversao e convencional. **Revista Brasileira de Armazenamento**, Viposa, MG, v.4, p.38-44, 2002.

THEODORO, V. C. A. de et al. Alterapoes quimicas em solo submetido a diferentes formas de manejo do cafeeiro. **Revista Brasileira de Ciencia do Solo**, v.27, p.1039-1047, 2003b.

THEODORO, V. C. A. de et al. Microbial biomass carbon and mycorrhizae in soil under native forest and coffee agroecosystems. **Acta Scientiarum**. Maringa, v.25, n.1, p.147-153, 2003a.

THEODORO, V. C. A. de et al. Avaliapao do estado nutricional de agroecossistemas de cafe organico no Estado de Minas Gerais. **Ciencia e Agrotecnologia.** Lavras, v.27, n.6, p.1222-1230, 2003c.

THEODORO, V. C. A. de. Certificapao de cafe organico. **Informe Agropecuario.** Belo Horizonte, v.23, n. 214/215, p.136-148, 2002.

THEODORO, V. C. A. de; GUIMARAES, R. J. O que significa cafe organico? **Cafeicultura**, Patrocinio, n.7, p.16-19, 2003.

THEODORO, V. C. de A. Certificapao de cafe organico. **Informe Agropecuario**, Belo Horizonte, v.23, n. 214/215, p. 136-148, 2002.

THEODORO, V. C. DE A.; CAIXETAS I. F.; GUIMARAES, R. J. **Bases para produpao de cafe organico**, MG. UFLA/PROEX, 2001.

TORRES, J. L. R.; PEREIRA, M. G.; JUNIOR RODRIGUES, D. J.; LOSS, A. Production, decomposition of residues and yield of maize and soybeans grown on cover crops. **Revista Ciencia Agronomica,** v. 46, n. 3, p. 460-468, 2015.

TOTE, A; P, ANDRADE, M; A **Educapao ambiental no centro estadual de educagao continuada** - CESEC- Betim, MG, December 2009.

TRIGUEIRO, R. M.; GUERRINI, I. A. Uso de biossolido como substrato para produgao de mudas de eucalipto. **Revista Scientia Florestalis**, v.64, n.2, p.150-162, 2003.

VAN RAIJ, B. **Fertilidade do solo e adubapao**. Ceres, Potafos, 1991, 343p.

VENTURA, J. A et al. **Diagnostico e manejo das doenenpas do cafeeiro conilon**. In: FERRAO, R. G. et al. (Eds). Cafe conilon. Vitoria, ES: Incaper, 450-497. 2007.

VENZON, M.; PALLINI, A.; AMARAL, D. S. S. L. Estrategias para o manejo ecologico de pragas. **Informe Agropecuario**, Belo Horizonte, v.22, n.212, p. 19-28, 2001.

VITAL, A. R. T.; GUERRINI, I. A.; FRANKEN, W. K.; FONSECA, R. C. B. Sapwood production and nutrient cycling in a semideciduous seasonal forest in riparian zone. **Revista Arvore,** v. 28, n. 6, p.793-800, 2004.

ZACARIAS, A. J.; PEREIRA, I. M; SOUZA, M. N.; LIMA, W. L; RANGEL, O. J. P. Efeito de adubos verdes em consorcio com cafeeiro e sua viabilidade econômica. **Resumos....**I Encontro Anual de Agroecologia e Qualidade de Vida do Ifes campus de Alegre. Poster and oral presentation. 2019.

ZACARIAS, A. J.; SOUZA, M. N. Recuperagao de area degradada de monocultura intensiva no estado do Espirito Santo. **REVISTA DA UNIVAP**, v.1, n. 87, p.234-242, 2019.

Printed by Books on Demand GmbH, Norderstedt / Germany